Circle the better choice.

1. You are more likely to pull a white cube than a gray cube.

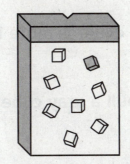

2. You are less likely to pull a gray cube than a black cube.

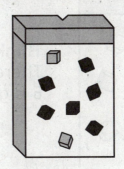
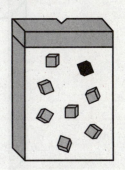

3. You are more likely to pull a gray cube than a white cube.

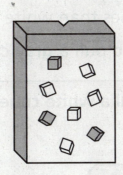
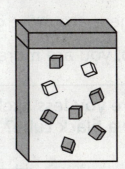

Check

4. Sybill says that if 7 more gray cubes were added to the bag below, she would be more likely to pull a gray cube than a black cube. Is she correct? Explain.

_____

_____

_____

_____

Name_____

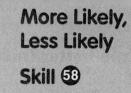
Learn the Math

You can tell whether an event is more likely or less likely.

Are you more likely to pull a gray cube or a white cube?

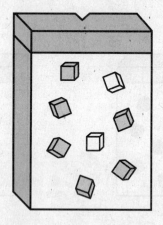

Think: There are 6 gray cubes.
There are 2 white cubes.
6 is greater than 2.

There are more gray cubes than white cubes.

So, you are more likely to pull a **gray** cube.

_____

Are you less likely to pull a white cube or a black cube?

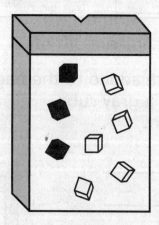

Think: There are 5 white cubes.
There are 3 black cubes.
3 is less than 5.

There are fewer black cubes than white cubes.

So, you are less likely to pull a _____ cube.

© Houghton Mifflin Harcourt

HOUGHTON MIFFLIN HARCOURT

Go Math!

Intensive Intervention

 **RtI** **Response to Intervention Tier 3**

Skill Packs

Grade 3

Houghton Mifflin Harcourt

ISBN 978-0-544-24916-5

1 2 3 4 5 6 7 8 9 10   0928   22 21 20 19 18 17 16 15 14 13

4500423464   A B C D E F G

Contents

Skills

1 Numbers to 20 .... IIN1

2 Tens IIN3

3 Tens and Ones IIN5

4 Understand Place Value IIN7

5 Compare Numbers to 50 .... IIN9

6 Order 2-Digit Numbers with Models IIN11

7 Count On to Add IIN13

8 Doubles and Doubles Plus One IIN15

9 Make a Ten IIN17

10 Add Tens IIN19

11 Count Back to Subtract IIN21

12 Think Addition to Subtract IIN23

13 Subtract Tens IIN25

14 Skip-Count by Fives and Tens IIN27

15 Pennies, Nickels, and Dimes IIN29

16 Quarters IIN31

17 Time to the Hour IIN33

18 Time to the Half Hour IIN35

19 Read a Tally Table IIN37

20 Picture Graphs IIN39

21 Bar Graphs IIN41

22 Describe Patterns IIN43

23 Extend Patterns IIN45

24 Skip-Count by 2s, 3s, 4s, 5s, and 10s IIN47

25 Skip-Count on a Hundred Chart IIN49

26 Skip-Count on a Number Line IIN51

27 Repeated Addition IIN53

28 Explore Multiplication IIN55

29	Use Models to Multiply	IIN57
30	Doubles and Multiplication	IIN59
31	Skip-Count to Multiply	IIN61
32	Arrays	IIN63
33	Multiply in Any Order	IIN65
34	Explore Equal Groups	IIN67
35	Make Equal Groups	IIN69
36	Size of Shares	IIN71
37	Number of Equal Shares	IIN73
38	Two Equal Groups	IIN75
39	Three Equal Groups	IIN77
40	Identify Plane Figures	IIN79
41	Classify Plane Figures	IIN81
42	Same Size, Same Shape	IIN83
43	Congruent Figures	IIN85
44	Identify Solid Figures	IIN87
45	Classify Solid Figures	IIN89
46	Use Nonstandard Units to Measure Length	IIN91
47	Inches	IIN93
48	Use Nonstandard Units to Measure Capacity	IIN95
49	Use Nonstandard Units to Measure Weight	IIN97
50	Centimeters	IIN99
51	Explore Perimeter	IIN101
52	Explore Area	IIN103
53	Equal Parts	IIN105
54	Halves	IIN107
55	Fourths	IIN109
56	Thirds	IIN111
57	Certain or Impossible	IIN113
58	More Likely, Less Likely	IIN115

Learn the Math

You can use ten frames to count, read, and write numbers to 20.

Remember
11 eleven
12 twelve
13 thirteen
14 fourteen
15 fifteen
16 sixteen
17 seventeen
18 eighteen
19 nineteen
20 twenty

Count 10 and 6 more.

Number: ___16___   Word: ___sixteen___

_____

Count 10 and 1 more.

Number: _____   Word: _____

_____

 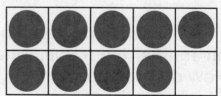

Count 10 and 9 more.

Number: _____   Word: _____

Do the Math

Circle the number word that tells how many.
Write the number.

1.  nineteen

   eighteen _____

2. 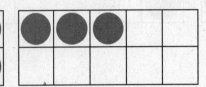 twelve

   thirteen _____

3. 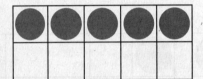 fifteen

   sixteen _____

4. 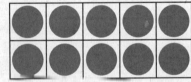 ten

   twenty _____

Check

5. Sara filled one ten frame with counters. Then she put 4 more counters in another ten frame. What number does she show? How do you know?

   _____

   _____

   _____

   _____

Learn the Math

You can group 10 ones to make a ten.

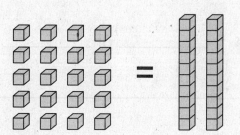

Vocabulary

ones
tens

Remember

10	ten
20	twenty
30	thirty
40	forty
50	fifty
60	sixty
70	seventy
80	eighty
90	ninety

10 ones = 1 ten

So, 1 ten = 10 or ten.

_____

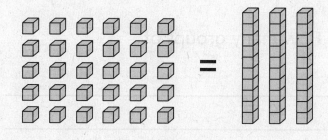

20 ones = _____ tens

So, 2 tens = _____ or _____.

_____

30 ones = _____ tens

So, 3 tens = _____ or _____.

Write how many tens. Write the number. Write the number word.

1.
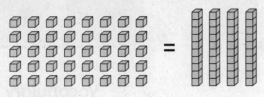

40 ones = _____ tens

_____ or _____

_____

2.
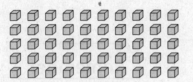

50 ones = _____ tens

_____ or _____

_____

3.

_____ tens = _____ or _____

_____

4.
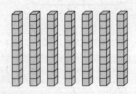

_____ tens = _____ or _____

5. Maria has a box of 60 crayons. How many groups of
ten can she make? Explain.

_____

_____

_____

_____

Learn the Math

You can use place value to write a number in different ways.

Tens	Ones

Vocabulary

ones
tens

Write the number as tens and ones.

_____ tens   _____ ones

Find the value of the tens and ones.

2 tens = _____   4 ones = _____

20 + _____

Add the values of the tens and ones.

20 + _____

24

Do the Math

Write each number three different ways.

1.

Tens	Ones

4 tens   7 ones

40 + _____

_____

___

2.

Tens	Ones

_____ tens _____ ones

30 + _____

_____

3.

Tens	Ones

_____ tens _____ ones

_____ + _____

_____

Check

4. Does the model represent 14 or 41? Explain.

Tens	Ones

_____

_____

_____

Name_____

Learn the Math

Each digit in a number has a value.

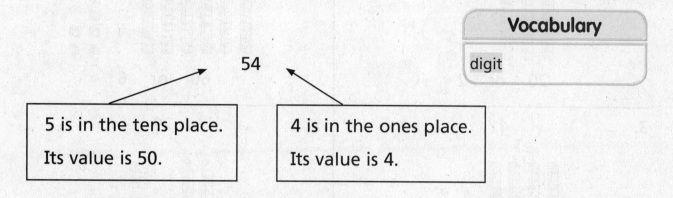

54

5 is in the tens place.	4 is in the ones place.
Its value is 50.	Its value is 4.

Vocabulary

digit

Circle the value of the underlined digit.

4<u>8</u>

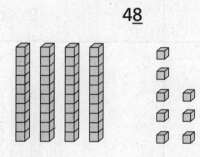

8 is in the ones place.

80   or   8

2<u>3</u>

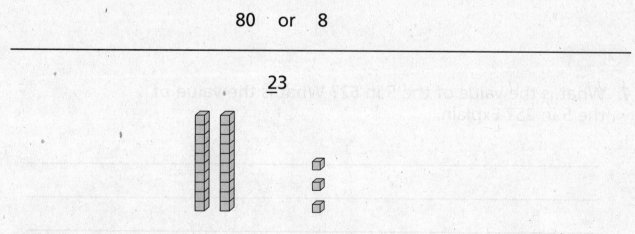

2 is in the tens place.

20   or   2

© Houghton Mifflin Harcourt

Circle the value of the underlined digit.

1.   3<u>2</u>

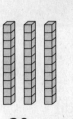

20   or   2

2.   <u>6</u>7

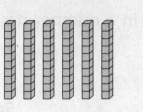

60   or   6

3.   <u>4</u>1

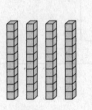

4   or   40

4.   2<u>9</u>

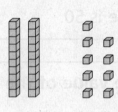

9   or   90

5.   <u>1</u>2

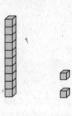

1   or   10

6.   4<u>6</u>

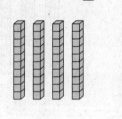

60   or   6

Check

7. What is the value of the 5 in 52? What is the value of the 5 in 25? Explain.

_____

_____

_____

_____

Learn the Math

You can use models and symbols to compare numbers.

Compare 39 and 31. Write <, >, or =.

Step 1 Model each number.

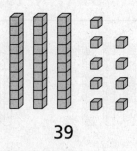
39

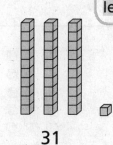
31

Step 2 Compare the tens.

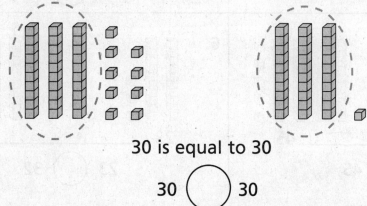

30 is equal to 30

30 ◯ 30

Step 3 Compare the ones.

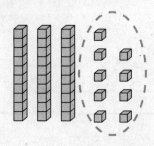   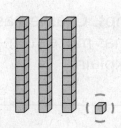

9 is greater than 1

9 ◯ 1

So, 39 ◯ 31.

Compare. Write <, >, or =.

1.

24 ___is less than___ 26

24 ◯ 26

2.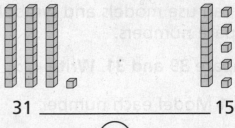

31 _____ 15

31 ◯ 15

3.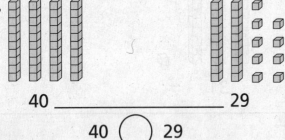

40 _____ 29

40 ◯ 29

4.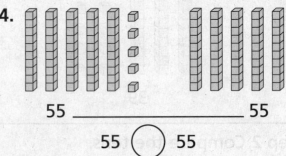

55 _____ 55

55 ◯ 55

5.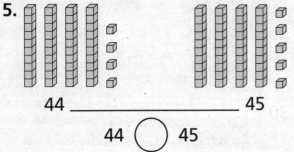

44 _____ 45

44 ◯ 45

6.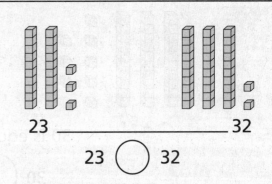

23 _____ 32

23 ◯ 32

Check

7. Lucy has 24 stamps. Charlie has 30 stamps.
Lucy thinks she has more stamps than Charlie.
Is she correct? Explain.

_____

_____

_____

_____

Learn the Math

You can use models and symbols to compare numbers.

You can order from least to greatest.

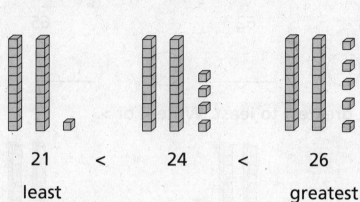

21   <   24   <   26

least   greatest

You can order from greatest to least.

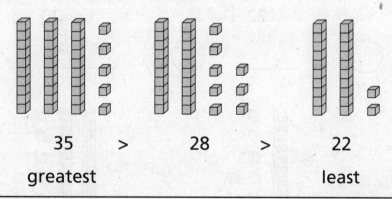

35   >   28   >   22

greatest   least

Write the numbers in order from greatest to least.
Then write < or >.

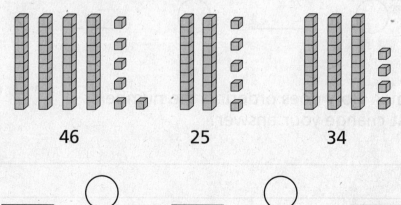

46   25   34

____ ◯ ____ ◯ ____

greatest   least

Order the numbers from least to greatest. Write < or >.

1.

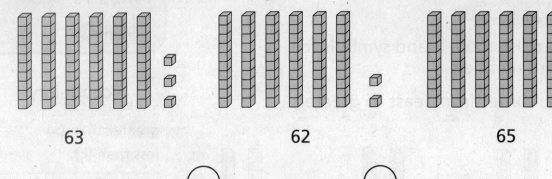

63   62   65

____ ◯ ____ ◯ ____

Order the numbers from greatest to least. Write < or >.

2.

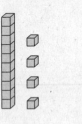   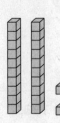   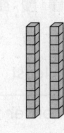

14   22   20

____ ◯ ____ ◯ ____

3.

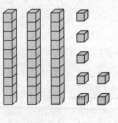   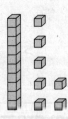   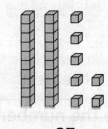

37   17   27

____ ◯ ____ ◯ ____

Check

4. Look back at Problem 3. How does ordering the numbers from least to greatest change your answer?

_____

_____

_____

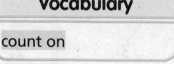

Learn the Math

You can use a number line to count on to add.

Add. 7 + 2 = __?__

Start with the greater addend.

Say 7. Move to the right 2 spaces.

Count **8, 9.**

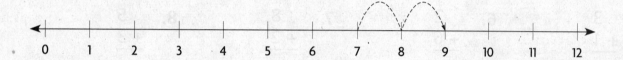

What number do you land on? _____

So, 7 + 2 = _____ .

_____

3 + 5 = __?__

Start with the greater addend.

Say 5. Move to the right 3 spaces.

Count **6, 7, 8.**

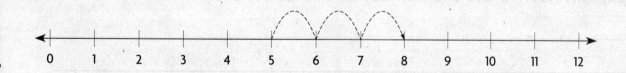

What number do you land on? _____

So, 3 + 5 = _____ .

Use the number line. Write the sum.

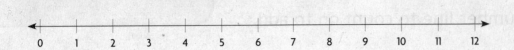

0 1 2 3 4 5 6 7 8 9 10 11 12

1. 3
    + 2

2. 3
    + 3

3. 1
    + 6

4. 9
    + 2

5. 3
    + 1

6. 2
    + 6

7. 8
    + 2

8. 5
    + 3

9. 1
    + 9

10. 2
    + 9

11. 7
    + 3

12. 3
    + 8

Check

13. Maya has 6 stickers in her book. She adds 3 more stickers. How many stickers does she have in all? Explain how to use the number line to count on.

_____

_____

_____

_____

Name_____

Learn the Math

You can use doubles and doubles plus one to help you find sums.

3 + 3 = ___?___

The addends are the same in a doubles fact.

So, 3 + 3 = _____ .

Use the doubles fact to find the doubles plus one fact.

3 + 4 = ___?___

3 + 4

4 is one more than 3. This is a doubles plus one fact.

Think: 3 + 3 plus 1 more.

3 + 3 = _____   So, 3 + 4 = _____ .

© Houghton Mifflin Harcourt

Write the doubles and doubles plus one facts.

1.

_____ + _____ = _____   _____ + _____ = _____

2.

_____ + _____ = _____   _____ + _____ = _____

3.

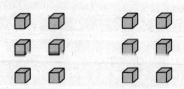   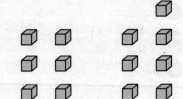

_____ + _____ = _____   _____ + _____ = _____

4.

_____ + _____ = _____   _____ + _____ = _____

Check

5. Liam delivered 6 newspapers and Azia delivered 7 newspapers. How many newspapers did they deliver in all?

_____

_____

_____

Name_____

Learn the Math

You can use a ten frame to find the sum.

Add. 7 + 5 = _?_

Step 1 Show 7 and 5.
Put 7 counters in the ten frame.
Put 5 counters below.

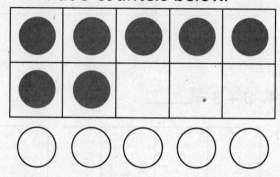

Step 2 Make a ten.
Move 3 counters into the ten frame.

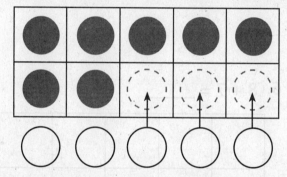

Step 3 Now you have ten and 2.

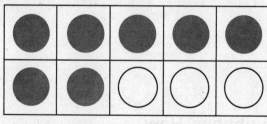

10 + 2 = _____   So, 7 + 5 = _____ .

© Houghton Mifflin Harcourt

Use the ten frame to find the sum.
Draw counters to show each addend.

1. 9 + 3 = _____

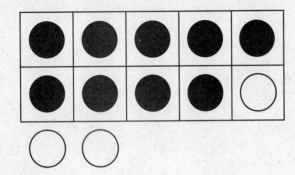

2. 7 + 4 = _____

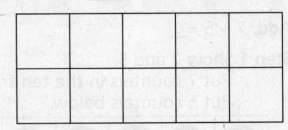

3. 8 + 5 = _____

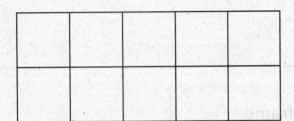

4. 8 + 3 = _____

5. 9 + 4 = _____

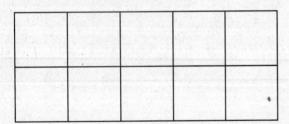

6. 8 + 7 = _____

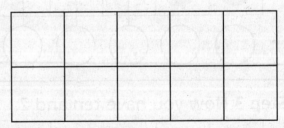

Check

7. Whitney has 9 red ribbons and 5 blue ribbons. How many ribbons does she have in all?

_____

Learn the Math

You can use mental math to add tens.

Add. 30 + 10 = _?_

Step 1 Start with the first addend.	How many tens are there in 30? ____ tens
Step 2 Look at the second addend.	How many tens are there in 10? ____ ten
Step 3 Add the number of tens.	What is the sum when you add the tens? ____ tens + ____ ten = ____ tens
Step 4 Find the value of the tens.	____ tens are equal to ____ .

So, 30 + 10 = _____ .

Add. 40 + 30 = _?_

40	+	30	=	_?_	
__4__ tens	+	__3__ tens	=	____ tens	
40	+	30	=	____	

Write how many tens. Then add.

1.   20   +   10   = __?__

___ tens + ___ ten = ___ tens

20   +   10   = ___

_____

2.   40   +   20   = __?__

___ tens + ___ tens = ___ tens

40   +   20   = ___

_____

3.   30   +   20   = __?__

___ tens + ___ tens = ___ tens

30   +   20   = ___

_____

4.   50   +   30   = __?__

___ tens + ___ tens = ___ tens

50   +   30   = ___

Check

5. What pattern do you see when you add tens?

_____

_____

_____

_____

_____

Name_____

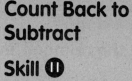

Learn the Math

You can use a number line to help you subtract.

Subtract. 7 − 3 = ___?___

Start at 7.

Move to the left 3 spaces.

Count **6, 5, 4**.

Vocabulary

count back

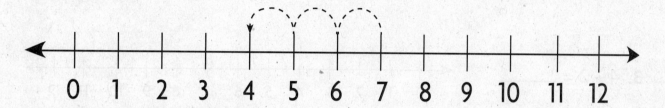

What number do you land on? ___4___

So, 7 − 3 = _____ .

_____

Subtract. 8 − 2 = ___?___

Start at 8.

Move to the left 2 spaces.

Count **7, 6**.

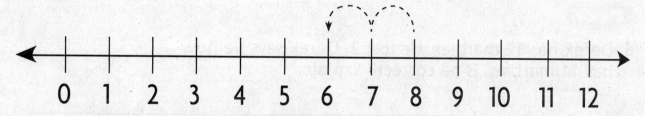

What number do you land on? _____

So, 8 − 2 = _____ .

Use the number line. Write the difference.

1. 6 – 2 = _____
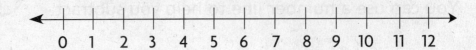

2. 9 – 1 = _____
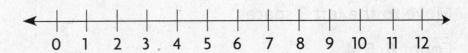

3. 4 – 2 = _____
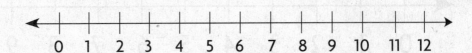

4. 7 – 2 = _____
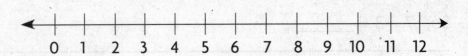

5. 3 – 2 = _____
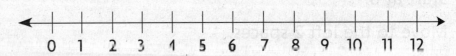

Check

6. Derek has 12 marbles. He lost 2. Derek says he now has 11 marbles. Is he correct? Explain.

_____

_____

_____

_____

Learn the Math

You can use addition to help you subtract.

Subtract. 7 – 4 = __?__

Think: 4 + __?__ = 7

4 + __3__ = 7

Since 4 + __3__ = 7, then 7 – 4 = __3__.

So, 7 – 4 = _____.

Vocabulary

subtract

_____

Subtract. 9 – 2 = __?__

Think: 2 + __?__ = 9

2 + __7__ = 9

Since 2 + __7__ = 9, then 9 – 2 = _____.

So, 9 – 2 = _____.

_____

Subtract. 5 – 3 = __?__

Think: 3 + __?__ = 5

3 + _____ = 5

Since 3 + _____ = 5, then 5 – 3 = _____.

So, 5 – 3 = _____.

Do the Math

Think addition to subtract.

1. $8 - 6 =$ ___?___

Since $6 +$ _____ $= 8$, then $8 - 6 =$ _____.

_____

2. $4 - 1 =$ ___?___

Since $1 +$ _____ $= 4$, then $4 - 1 =$ _____.

_____

3. $9 - 3 =$ ___?___

Since $3 +$ _____ $= 9$, then $9 - 3 =$ _____.

_____

4. $7 - 6 =$ ___?___

Since $6 +$ _____ $= 7$, then $7 - 6 =$ _____.

_____

5. $8 - 4 =$ ___?___

Since $4 +$ _____ $= 8$, then $8 - 4 =$ _____.

Check

6. James had 6 toy cars. He lost 2 cars. What addition
   fact can you use to find how many cars he has
   now? Explain.

   _____

   _____

   _____

Learn the Math

You can use mental math to subtract tens.

40 − 20 = _?_

Vocabulary

tens

Step 1

Start with the first number.

How many tens are there in 40?

___4___ tens

Step 2

Look at the second number.

How many tens are there in 20?

___2___ tens

Step 3

Subtract the number of tens.

What is the difference?

_____ tens − _____ tens = _____ tens

Step 4

Find the value of the tens.

_____ tens is equal to _____.

So, 40 − 20 = _____.

Do the Math

Write how many tens. Then subtract.

1.   30   –   10   = __?__

    ___ tens – ___ ten   = ___ tens

    30   –   10   = ___

2.   60   –   40   = __?__

    ___ tens – ___ tens   = ___ tens

    60   –   40   = ___

3.   50   –   10   = __?__

    ___ tens – ___ ten   = ___ tens

    50   –   10   = ___

4.   40   –   30   = __?__

    ___ tens – ___ tens   = ___ ten

    40   –   30   = ___

Check

5. There are 60 crayons in a box. Shari used 30 of the crayons. How many crayons did Shari not use?

_____

_____

_____

Name_____

Learn the Math

You skip-count by fives and tens to solve problems.

Vocabulary

skip-count

Skip-count by fives to find how many apples.

 5   10   15   20   _____ apples

Skip-count by tens to find the missing numbers.

 10   _____   30   _____   50

Skip-count. Write the missing numbers.

5, 10, _____, 20, _____, 30, 35, _____, _____, 50

10, _____, _____, 40, 50, _____, 70, _____

Skip-count by fives. Write the missing numbers.

1.   40, 45, _____, _____, 60, 65, _____, _____

2.   25, _____, 35, _____, 45, _____, 55, _____, 65

3.   10, 15, _____, _____, 30, 35, _____, 45, _____, _____, 60

Skip-count by tens. Write the missing numbers.

4.   _____, 40, 50, _____, 70 , _____, _____

5.   20, _____, _____, 50, 60, _____

6.   _____, 20, 30, 40, _____, 60, 70, _____, _____

Check

7. There are 5 eggs in each nest. How many eggs are there in 6 nests? Explain how to use skip-counting to find the answer.

_____

_____

_____

Name_____

Learn the Math

You can count by tens, fives, and ones to find the total value of the coins.

Step 1
Count by tens.

10¢,   20¢,   __?__ ,   __?__ ,   __?__

Step 2
Count by fives.

10¢,   20¢,   25¢,   30¢,   __?__

Step 3
Count by ones.

10¢,   20¢,   25¢,   30¢,   31¢

So, 2 dimes, 2 nickels, and 1 penny is _____ .

Count by tens, fives or ones. Write the total value.

1.

_____ ¢   _____ ¢   _____ ¢   _____ ¢   _____ ¢

2.

_____ ¢   _____ ¢   _____ ¢   _____ ¢

3.

_____ ¢   _____ ¢   _____ ¢   _____ ¢

Check

4. Ana has a dime and a nickel. She says that she has 11¢.
Is she correct? Explain.

_____

_____

_____

_____

Name _____

Learn the Math

You can skip-count to find the total value of a collection of coins.

 or    1 quarter = 25 cents

---

Find the total value of the coins.

Step 1 Group the coins in order from greatest to least value.

Step 2 Skip-count to find the total value.
Count by 25s for quarters. Count by 10s for dimes.
Count by 5s for nickels. Count by 1s for pennies.

     25¢,   50¢,   _____ ¢,   _____ ¢,   66¢

So, the total value of the coins is _____ ¢.

Count the coins to find the total value.

1.

Total

_____ ¢

_____ ¢, _____ ¢, _____ ¢, _____ ¢, _____ ¢

2.

Total

_____ ¢

_____ ¢, _____ ¢, _____ ¢, _____ ¢, _____ ¢

3.

Total

_____ ¢

_____ ¢, _____ ¢, _____ ¢, _____ ¢, _____ ¢

4.

Total

_____ ¢

_____ ¢, _____ ¢, _____ ¢, _____ ¢, _____ ¢

Check

5. Sasha has 3 dimes, 2 quarters, and 1 nickel. How can she order
the coins to count them? What is the total value of her coins?

_____

_____

_____

Name_____

Learn the Math

A clock shows us the time.

Show 5:00 on the clock.

Step 1
Start with the minute hand and the hour hand pointing to 12.

Vocabulary

hour
hour hand
minute
minute hand

Step 2
Move the hour hand to the 5.

Minute Hand   Hour Hand

You can also show the time on a digital clock.

Both clocks show ___:___.

Use the clock to read each time.
Write the time.

1.

```
 :
```

2.

```
 :
```

3.

```
 :
```

4.

```
 :
```

5.

```
 :
```

6.

```
 :
```

Draw the clock hands.

7.

8:00

8.

10:00

9.

6:00

Check

10. Where is the minute hand if the time is 4 o'clock?
Where is the hour hand?

_____

_____

_____

_____

Learn the Math

Lunch begins at 11:00 and ends a half hour later. What time does lunch end?

Step 1
Start with the minute hand at 12 and the hour hand at 11.

Step 2
A half an hour is 30 minutes.

Move the minute hand to the 6 and the hour hand halfway between the 11 and 12.

Lunch ends at _____ .

You can show the time on a digital clock.

```
11:30
```

© Houghton Mifflin Harcourt

Do the Math

Write the time.

1.

2.

3.

4.

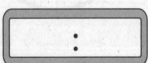

5.

6.

Check

7. Elizabeth started her homework at 4:00. She finished half an hour later. What time did Elizabeth finish her homework? Where would the minute hand be on the clock?

_____

_____

_____

Name_____

Learn the Math

You can use a tally table to show information.

Football Games Jacob Played	
Month	**Tally**
September	ⅢⅡ I
October	I I
November	I I I I

Vocabulary

tally table
tally mark

Each tally mark I stands for 1 game.

ⅢⅡ stands for 5 games.

In which month did Jacob play the most football games?

Step 1 Look at the tally table. Count the tally marks to find the total number of games played each month.

ⅢⅡ I = __6__ games in September

I I = ____ games in October

I I I I = ____ games in November

Step 2 Compare the numbers of games played each month.

Jacob played 6 games in September, 2 games in October, and 4 games in November.

6 (>) 2   6 ( ) 4

So, Jacob played the most games in September.

© Houghton Mifflin Harcourt

Use the tally table to answer each question.

Evan brought muffins to class. He brought three different flavors.

Muffin Flavors	
Flavor	**Tally**
blueberry	⅏
banana	⅏⅏ ||||
bran	|||

1. How many blueberry muffins did Evan bring? _____ blueberry

2. How many banana muffins did Evan bring? _____ banana

3. How many bran muffins did Evan bring? _____ bran

4. Which kind of muffin did Evan bring the most of? _____

5. Which kind of muffin did Evan bring the fewest of? _____

6. How many muffins did Evan bring in all? _____ muffins

Check

7. Look at the tally table above. How many more blueberry muffins are needed for an equal number of blueberry and banana muffins? How would you show this information?

_____

_____

_____

© Houghton Mifflin Harcourt

Learn the Math

A picture graph uses pictures to show information.

This picture graph shows the favorite sports of Ian's friends.

Our Favorite Sports

baseball ⚾	⚾	⚾	⚾	⚾		
basketball 🏀	🏀	🏀	🏀			
soccer ⚽	⚽	⚽	⚽	⚽	⚽	⚽

Which sport did the most students choose?

Step 1 Look at the number of pictures in each row.
How many students chose each sport?

baseball = __4__ students

basketball = ____ students

soccer = ____ students

Step 2 Compare the numbers to see which is the greatest.

6 ◯ 4 ◯ 3

So, _____ is the sport that the most students chose.

Do the Math

Use the picture graph to answer the questions.

Our Favorite Fruits

apple								
peach								
banana								

1. How many students chose apple? _____ students

2. How many students chose peach? _____ students

3. How many students chose banana? _____ students

4. Which fruit did the fewest students choose? _____

5. Which fruit did the most students choose? _____

Check

6. How many students chose apples or bananas as their favorite fruit? How did you find your answer?

_____

_____

_____

_____

Name_____

Learn the Math

A bar graph uses bars to show amounts.

Mr. Smith asked his students which subject they liked best. He made a bar graph to show the results.

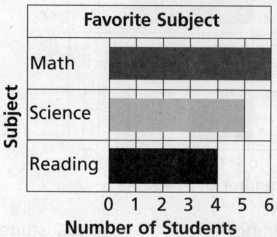

Look at where the bars end. This tells how many.

How many students did Mr. Smith ask?

Step 1 Find the number of students who chose each subject.

Math   = ___6___ students

Science   = _____ students

Reading   = _____ students

Step 2 Add to find the total number of students.

___6___ + _____ + _____ = _____

So, Mr. Smith asked _____ students.

Which subject did the most students choose?
Compare the lengths of the bars on the graph.

Which bar is the longest? _____

So, the most students chose _____ .

Use the bar graph to answer the questions.

Aimee made a bar graph to show her classmates' favorite kinds of juice.

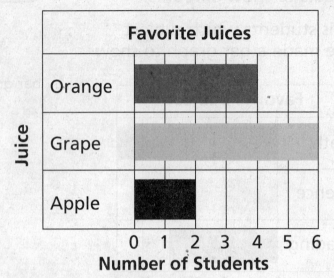

1. How many students chose orange juice?   _____ students

2. How many students chose grape juice?   _____ students

3. How many students chose apple juice?   _____ students

4. Which juice did the most students choose?   _____

5. Which juice did the fewest students choose?   _____

6. How many classmates did Aimee ask?   _____ classmates

Check

7. Look at the bar graph above. Write the juices in order from least favorite to most favorite.

_____

_____

Name_____

Learn the Math

A pattern unit repeats over and over to make a repeating pattern.

This example shows a repeating pattern of a white circle and a gray circle.

Vocabulary
pattern unit
repeating pattern

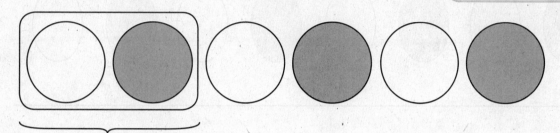

pattern unit

This repeating pattern shows two white triangles and one gray triangle.

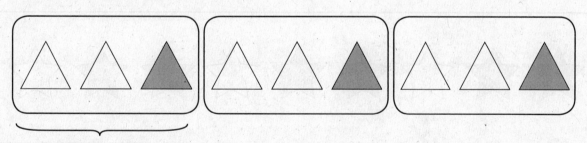

pattern unit

Circle each pattern unit. Describe the pattern.

The pattern unit is _____

_____.

Circle each pattern unit. Describe the pattern.

1.

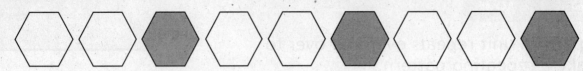

_____

2.

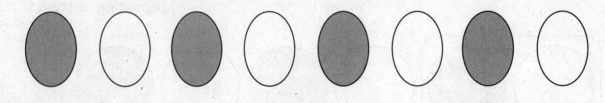

_____

3.

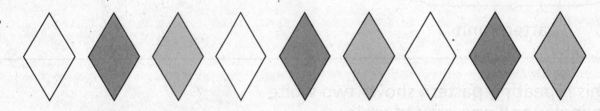

_____

4.

_____

5. Hayden says the pattern unit for the pattern below is white square, gray square, white square. Is he correct? Explain. Then circle each pattern unit.

_____

_____

_____

Name_____

Learn the Math

Look at the pattern unit to help you find what comes next.

The pattern unit is circle, triangle, triangle.
A circle comes next.

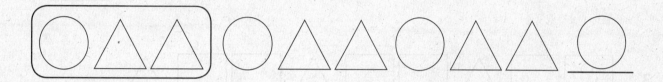

The pattern unit is square, square, star.
A square comes next.

Circle the pattern unit. Draw to continue the pattern.

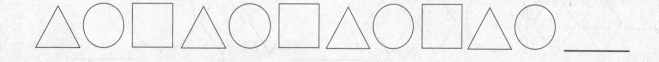

Circle the pattern unit. Draw to continue the pattern.

1.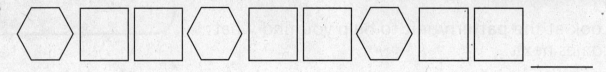

2.

3.

4.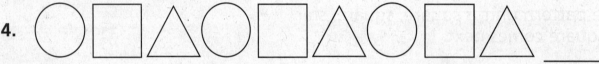

5.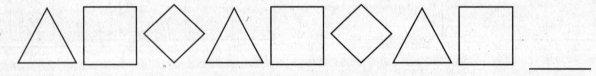

6. Alex drew this pattern. What is missing? How do you know? Draw the missing figures.

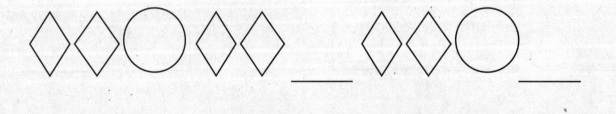

_____

_____

_____

Learn the Math

There are 2 wheels on a bicycle. How many wheels are on 4 bicycles?

Skip-count to find how many.

Step 1 Use pictures or objects.

Step 2 Skip-count by twos.

  __2__   __4__   ____   ____

Step 3 Write the answer.

_____ wheels

So, 4 bicycles have 8 wheels.

Skip-count. Write how many.

1.

_____ _____ _____

_____ balloons

2.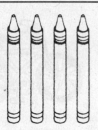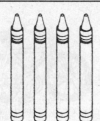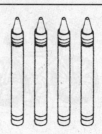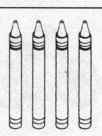

_____ _____ _____ _____

_____ crayons

3.

_____ _____ _____ _____

_____ books

Check

4. Tyler bought 6 packs of baseball cards. Each pack has 10 cards in it. How many cards did Tyler buy in all?

_____ cards

Learn the Math

Skip-count by fives using a hundred chart.

Step 1

Circle the number 5.
Shade the square.

Vocabulary

skip-count

Step 2

Count by fives.
Shade the squares.

1	2	3	4	5	6	7	8	9	10
11	12	13	14	15	16	17	18	19	20
21	22	23	24	25	26	27	28	29	30
31	32	33	34	35	36	37	38	39	40
41	42	43	44	45	46	47	48	49	50
51	52	53	54	55	56	57	58	59	60
61	62	63	64	65	66	67	68	69	70
71	72	73	74	75	76	77	78	79	80
81	82	83	84	85	86	87	88	89	90
91	92	93	94	95	96	97	98	99	100

Step 3

Skip-count by fives.

5, 10, 15, 20, 25, 30, 35, 40, _____ , _____ , 55, 60,

65, _____ , _____ , _____ , 85, 90, _____ , _____

Skip-count. Show the pattern on the hundred chart.

1. Skip-count by tens.

1	2	3	4	5	6	7	8	9	10
11	12	13	14	15	16	17	18	19	20
21	22	23	24	25	26	27	28	29	30
31	32	33	34	35	36	37	38	39	40
41	42	43	44	45	46	47	48	49	50
51	52	53	54	55	56	57	58	59	60
61	62	63	64	65	66	67	68	69	70
71	72	73	74	75	76	77	78	79	80
81	82	83	84	85	86	87	88	89	90
91	92	93	94	95	96	97	98	99	100

2. Skip-count by fives.

1	2	3	4	5	6	7	8	9	10
11	12	13	14	15	16	17	18	19	20
21	22	23	24	25	26	27	28	29	30
31	32	33	34	35	36	37	38	39	40
41	42	43	44	45	46	47	48	49	50
51	52	53	54	55	56	57	58	59	60
61	62	63	64	65	66	67	68	69	70
71	72	73	74	75	76	77	78	79	80
81	82	83	84	85	86	87	88	89	90
91	92	93	94	95	96	97	98	99	100

Use the hundred chart to skip-count.

3. Skip-count by twos.

46, 48, 50, _____ , _____ , _____ , _____ , _____

4. Skip-count by threes.

12, 15, 18, _____ , _____ , _____ , _____ , _____

Check

5. Juan skip-counted by fours and shaded every number in the hundred chart that ended in 4. Is this correct? Explain.

_____

_____

_____

_____

Name_____

Learn the Math

You can use a number line to help you skip-count and find patterns.

Use the number line to skip-count. Write a rule for the pattern.

Vocabulary
pattern
skip-count

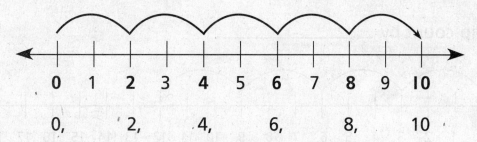

0,   2,   4,   6,   8,   10

Each number is 2 more than the number before it.

Rule: Skip-count by twos.

Find the missing numbers in the pattern. Write a rule.

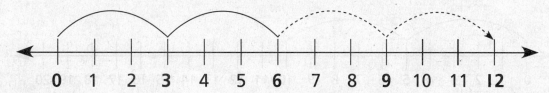

Think: Each number is 3 more than the number before it.

0,   3,   6, ____ , ____

Rule: Skip-count by _____ .

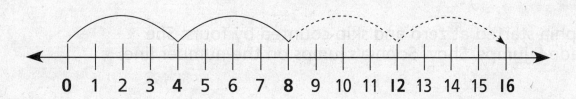

0,   4,   8, ____ , ____

Rule: Skip-count by _____ .

© Houghton Mifflin Harcourt

Skip-count. Draw the missing jumps and write the missing numbers. Write a rule for the pattern.

1.

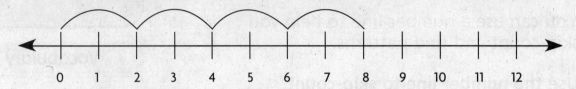

0,   2,   4,   6,   8, _____, _____

Rule: Skip-count by _____.

2.

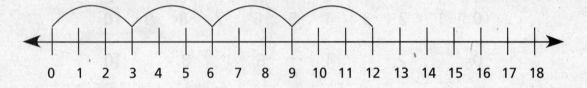

0,   3,   6,   9,   12, _____, _____

Rule: Skip-count by _____.

3.

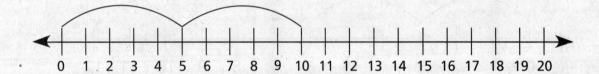

0,   5,   10, _____, _____

Rule: Skip-count by _____.

4. Sophia started at zero and skip-counted by fours. She made 5 jumps. Show Sophia's jumps on the number line.

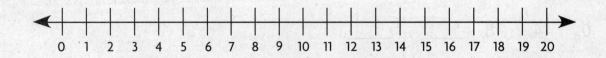

Learn the Math

When there are equal groups, you can use repeated addition to find how many in all.

Use repeated addition to find how many in all.

Step 1
Count the number of groups.

There are _____ groups.

Step 2
Count how many there are in each group.

There are _____ in each group.

4   4   4

Step 3
Add the equal groups to find how many in all.

_____ + _____ + _____ = _____

_____ groups of _____

So, 4 + 4 + 4 = _____ .

Do the Math

Write the number in each group and the number of groups.
Use repeated addition to find how many in all.

1. ____ groups of ____

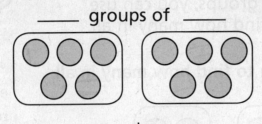

____ + ____ = ____

2. ____ groups of ____

____ + ____ + ____ = ____

3. ____ groups of ____

____ + ____ + ____ + ____ = ____

4 ____ groups of ____

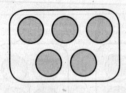 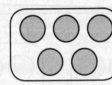 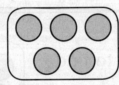 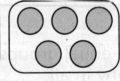

____ + ____ + ____ + ____ = ____

Check

5. Jonas says that 3 groups of 3 equals 6 because 3 + 3 = 6. What mistake did Jonas make?

_____

_____

Name_____

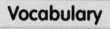

Learn the Math

When you multiply, you add equal groups.

4 groups of 2

Vocabulary

multiplication sentence
multiply

2   +   2   +   2   +   2   =   8

4   (×)   2   (=)   8

Write the sum and the multiplication sentence.

3 groups of 3

   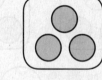

3   +   3   +   3   =   ____

____ ( ) ____ ( ) ____

---

4 groups of 4

4   +   4   +   4   +   4   =   ____

____ ( ) ____ ( ) ____

© Houghton Mifflin Harcourt

Do the Math

Write an addition sentence to find how many in all.
Then write a multiplication sentence.

1.   2 groups of 4

_____ + _____ = _____   _____ ◯ _____ ◯ _____

2.   3 groups of 5

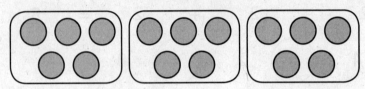

_____ + _____ + _____ = _____   _____ ◯ _____ ◯ _____

3.   4 groups of 3

_____ + _____ + _____ + _____ = _____   _____ ◯ _____ ◯ _____

Check

4. Jeremy wrote 6 × 6 = 12 for the model below. What
mistake did he make?

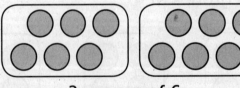

2 groups of 6

_____

_____

_____

Name_____

Learn the Math

You can use models to multiply.

Use the counters to multiply.

Vocabulary

factor
multiply
product

2 groups
3 in each group

$$2 \quad \times \quad 3 \quad = \quad 6$$

factor   product

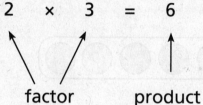

So, there are **6** counters in all.

Use the counters to multiply.

4 groups
2 in each group

$$4 \quad \times \quad 2 \quad = \quad 8$$

factor   product

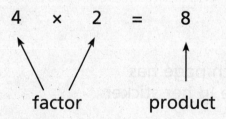

So, there are _____ counters in all.

© Houghton Mifflin Harcourt

Do the Math

Use counters to multiply.

1.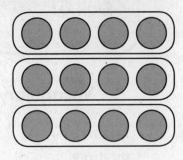

    3 × 4 = _____

2.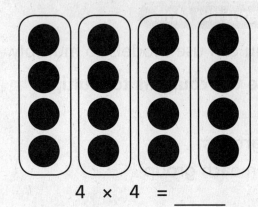

    4 × 4 = _____

3.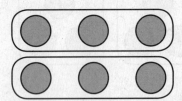

    2 × 3 = _____

4.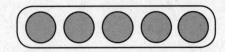

    1 × 5 = _____

5.

    3 × 3 = _____

6.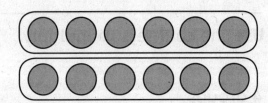

    2 × 6 = _____

Check

7. There are 3 pages in Violet's sticker book. Each page has
   5 stickers. How many stickers does Violet have in her sticker
   book? Explain how you solved the problem.

   _____

   _____

   _____

Name_____

Doubles and Multiplication

Skill ③⓪

Learn the Math

Doubling an addend is the same as multiplying by 2.

Find the product.
2 × 4

Use doubles.

You can multiply by 2 by doubling the other factor.

4 × 2 = 2 × 4 = 4 + 4 = _____

So, 2 × 4 equals 8.

Vocabulary

double
multiply

Use a model to multiply.

There are 2 groups.
There are 4 counters in each group.

2 groups of 4

So, 2 × 4 equals 8.

2 × 4 = _____

Find the product.
2 × 3

Use doubles.

3 × 2 = 2 × _____ = _____ + _____ = _____

Use a model.

2 groups of 3

_____ × _____ = _____

So, 2 × 3 equals _____ .

© Houghton Mifflin Harcourt

Use doubles to find the product.

1. 2 × 2 = _____

2 + 2 = _____

2. 2 × 3 = _____

3 + 3 = _____

Use a model to find the product.

3. 2 × 5 = _____

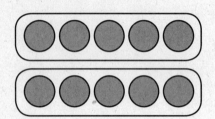

_____ groups of _____

2 × 5 = _____

4. 2 × 4 = _____

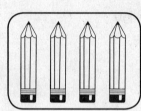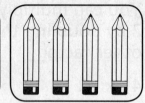

_____ groups of _____

2 × 4 = _____

5. 2 × 3 = _____

_____ groups of _____

2 × 3 = _____

6. 2 × 6 = _____

_____ groups of _____

2 × 6 = _____

7. Ben has 2 piles of baseball cards. Each pile has 5 cards. How many baseball cards does Ben have in all? Explain how to solve the problem.

_____

_____

Learn the Math

There are 3 tables. There are
4 children at each table.
How many children are there in all?

You can skip-count to find the product.

3 × 4 = ▪

Model

Use a number line.
Start at 0.
Skip-count by fours 3 times.

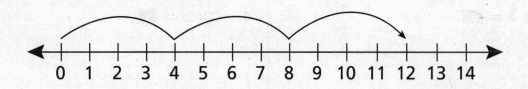

Think

Skip-count by 4s.

___4___ , ___8___ , _____

Record

Write the problem as a multiplication sentence.

3 × 4 = _____

So, there are _____ children in all.

Skip-count to find the product.
Then write the multiplication sentence.

1. $4 \times 2 =$ ■

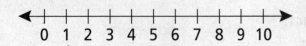

_____, _____, _____, _____

_____ × _____ = _____

2. $3 \times 3 =$ ■

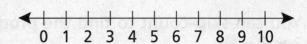

_____, _____, _____

_____ × _____ = _____

3. $5 \times 2 =$ ■

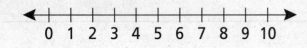

_____, _____, _____, _____, _____

_____ × _____ = _____

4. $2 \times 5 =$ ■

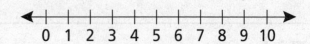

_____, _____

_____ × _____ = _____

Check

5. Annie has 3 bags of apples. There are 5 apples in each bag. How many apples does Annie have in all? Explain. Use the number line to skip-count.

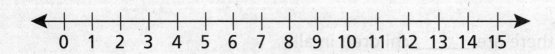

_____

_____

Learn the Math

You can use an array to model multiplication.

Write a multiplication sentence for the array.

row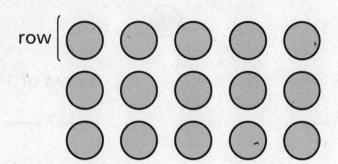

© Houghton Mifflin Harcourt

Vocabulary

array

Step 1
Count the number of rows.

_____ rows

Step 2
Count the number of counters in each row.

_____ in each row

Step 3
Write a multiplication sentence.

3 rows of 5

___3___ × _____ = _____

There are _____ counters in this array, so 3 × 5 equals _____.

Write the multiplication sentence for each array.

1.

2 rows of 5

____ × ____ = ____

2.

3 rows of 1

____ × ____ = ____

3. X X X X
X X X X
X X X X

3 rows of 4

____ × ____ = ____

4. X X X
X X X

2 rows of 3

____ × ____ = ____

5.

4 rows of 2

____ × ____ = ____

6.

3 rows of 3

____ × ____ = ____

Check

7. Mr. Williams' classroom has 5 rows of desks. There are 4 desks in each row. How many desks are in Mr. Williams' classroom? What multiplication sentence can you write?

_____

_____

Learn the Math

You can multipily numbers in any order. The product is the same.

Show that 2 × 4 and 4 × 2 have the same product.

Vocabulary

array
product

Step 1

Make an array to show 2 × 4.

_____ rows of _____

Step 2

Write the multiplication sentence.

2 × 4 = _____

Step 3

Turn the array to show 4 × 2.

_____ rows of _____

Step 4

Write the multiplication sentence.

_____ × _____ = _____

So, 2 × 4 = _____ and 4 × 2 = _____ .

Write the multiplication sentence for each array.

1.

5 rows of 2

____ × ____ = ____

2 rows of 5

____ × ____ = ____

2.

2 rows of 3

____ × ____ = ____

3 rows of 2

____ × ____ = ____

3.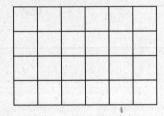

4 rows of 6

____ × ____ = ____

6 rows of 4

____ × ____ = ____

Check

4. Terry has 3 rows of dominoes. There are 8 dominoes in each row. He found the product of 8 × 3. Is this correct? Explain.

_____

_____

Name_____

Learn the Math

You can separate a set into equal groups
to find how many in each group.

Vocabulary

equal groups

Separate 10 counters into 2 equal groups.

Step 1 Use 10 counters.

Step 2 Show 2 groups.
Put one counter in each group.
Continue until all 10 counters are used.

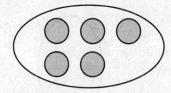

The groups are equal.
There are ___ counters in each group.

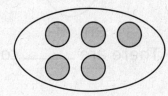

So when you separate 10 counters into 2 equal
groups, there are 5 counters in each group.

Separate 12 counters into 3 equal groups.

Write how many in each group.

There are ___ counters in each group.

So, when you separate 12 counters into 3 equal groups,
there are 4 counters in each group.

© Houghton Mifflin Harcourt

Find how many in each equal group.
Draw to show your work.

1. Separate 15 counters into 3 equal groups.

   There are _____ counters in each group.

2. Separate 8 counters into 4 equal groups.

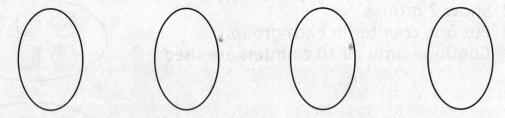

   There are _____ counters in each group.

3. Separate 10 counters into 5 equal groups.

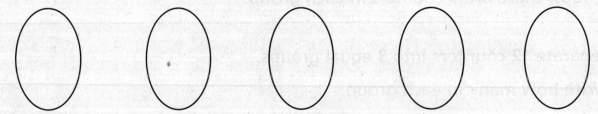

   There are _____ counters in each group.

Check

4. Look at Problem 3. How did you know how many
   counters to put in each group?

   _____

   _____

   _____

Learn the Math

You can separate a set by putting the same amount in each group to find the number of equal groups.

Separate 12 counters into groups of 4.

Vocabulary

equal groups

Step 1 Use 12 counters.

Step 2 Circle groups of 4 counters.

Step 3 Count the number of equal groups.
There are _____ equal groups.

So, when you separate 12 counters into groups of 4, there are 3 equal groups.

Circle equal groups. Write how many groups there are.
Separate 15 triangles into groups of 5.

So, when you separate 15 triangles into groups of 5, there are _____ equal groups.

Circle equal groups. Write how many groups there are.

1. Separate 9 counters into groups of 3.

   There are _____ equal groups.

2. Separate 18 stars into groups of 6.

   There are _____ equal groups.

3. Separate 20 moons into groups of 4.

   There are _____ equal groups.

Check

4. Ana has 14 game pieces for a game. Each player gets
   2 pieces. Can 8 people play Ana's game? Explain.

   _____

   _____

Name_____

Learn the Math

When you divide, you separate into equal groups.

Divide 6 counters into 2 equal groups.

Vocabulary	

divide
equal groups

Step 1
Use 6 counters.

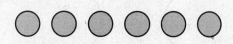

Step 2
Show 2 groups.
Place one counter in
each group.

Continue until all
6 counters are used.

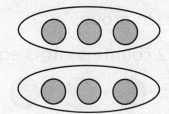

Step 3
Count how many counters
are in each group.

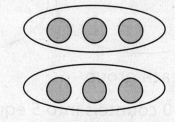

____ in each group

So, there are 3 counters in each group.

Divide 8 counters into 4 equal groups.
Draw to show your work.

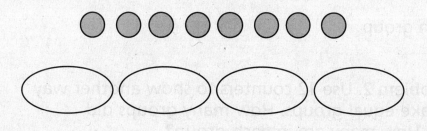

____ in each group

So, there are ____ counters in each group.

Divide. Draw to show your work.
Write how many in each group.

1. Divide 8 counters into 2 equal groups.

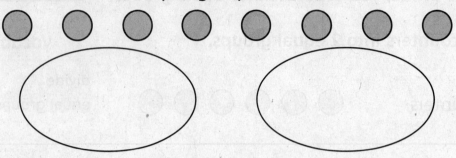

____ in each group

2. Divide 12 counters into 3 equal groups.

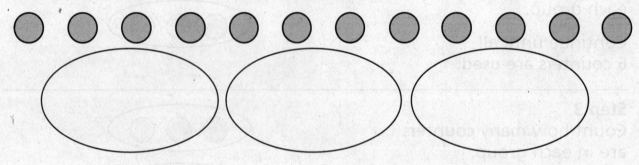

____ in each group

3. Divide 10 counters into 5 equal groups.

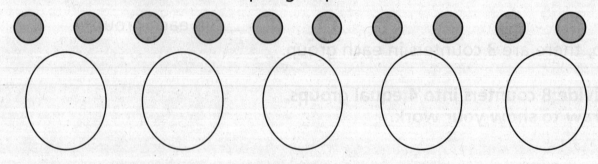

____ in each group

Check

4. Look at Problem 2. Use 12 counters to show another way you can make equal groups. How many groups did you make? How many are in each group?

_____

_____

Name_____

Learn the Math

When you divide, you put the same amount in each group to find the number of groups.

Divide 12 counters into groups of 6. How many equal groups are there?

Step 1
Use 12 counters.

Step 2
Circle as many groups of 6 as possible.

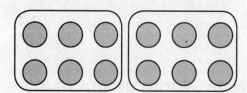

Step 3
Count the number of groups.

There are ___ equal groups.

So, 12 divided into groups of 6 makes ___ groups.

Circle equal groups. Write how many equal groups there are.

Divide 12 into groups of 3.

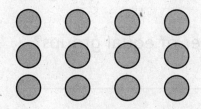

___ equal groups

Divide 16 into groups of 4.

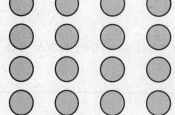

___ equal groups

Do the Math

Use counters to make equal groups. Draw to show your work. Write how many equal groups.

1. Divide 4 into groups of 2.

_____ equal groups

2. Divide 10 into groups of 5.

_____ equal groups

3. Divide 12 into groups of 4.

_____ equal groups

4. Divide 9 into groups of 3.

_____ equal groups

Check

5. When you divide, how do you find the number of equal groups?

_____

_____

_____

Name_____

Two Equal Groups
Skill 38

Learn the Math

You can divide any even number into two equal groups. You can write a division sentence to show how you divided.

Divide 12 into 2 equal groups. Write a division sentence.

Vocabulary

divide

division sentence

Step 1
Use 12 counters.

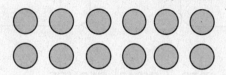

_____ in all

Step 2
Show 2 groups.

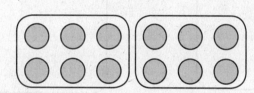

_____ equal groups

Step 3
Count how many counters in each group.

_____ in each group

So, 12 divided into 2 equal groups is _____.

Write: $12 \div 2 = 6$

© Houghton Mifflin Harcourt

Use counters to make equal groups. Draw the counters.
Write the numbers. Complete the division sentence.

1. Divide 8 into 2 equal groups.

_____ in all

_____ equal groups

_____ in each group

8 ÷ 2 = _____

2. Divide 6 into 2 equal groups.

_____ in all

_____ equal groups

_____ in each group

6 ÷ 2 = _____

3. Divide 10 into 2 equal groups.

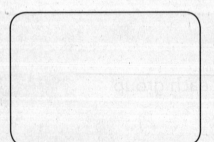

_____ in all

_____ equal groups

_____ in each group

10 ÷ 2 = _____

Check

4. Sam has 14 stickers. He put an equal number of stickers
on 2 pages. Write a division sentence to show how many
stickers he puts on each page.

Name_____

Learn the Math

You can write a division sentence to show how to divide a number into 3 equal groups.

Divide 6 into 3 equal groups.
Write a division sentence.

Vocabulary

divide
division sentence

Step 1
Use 6 counters.

_____ in all

Step 2
Show 3 groups.

_____ equal groups

Step 3
Count how many in each group.

_____ in each group

So, 6 divided into 3 equal groups is _____ .

Write: 6 ÷ 3 = 2

Use counters to make equal groups. Draw the counters.
Write the numbers. Complete the division sentence.

1. Divide 9 into 3 equal groups.

_____ in all

_____ equal groups

_____ in each group

9 ÷ 3 = _____

2. Divide 12 into 3 equal groups.

_____ in all

_____ equal groups

_____ in each group

12 ÷ 3 = _____

3. Divide 3 into 3 equal groups.

_____ in all

_____ equal groups

_____ in each group

3 ÷ 3 = _____

Check

4. Claire walks the same number of blocks to school each
day. She walked a total of 6 blocks in 3 days. Write a
division sentence to show how many blocks she walked
each day.

_____ _____ ◯ _____

© Houghton Mifflin Harcourt

Learn the Math

You can describe a plane figure in terms of its sides and vertices.

This figure has **3** sides and **3** vertices.

It is a triangle.

More Triangles

This figure has **4** sides and **4** vertices.

It is a rectangle.

More Rectangles

This figure has ___ sides

and ___ vertices.

It is a square.

More Squares

Remember: A square is a rectangle with all sides equal.

© Houghton Mifflin Harcourt

Color circles blue, triangles red, rectangles green, and squares yellow.

1.

2.
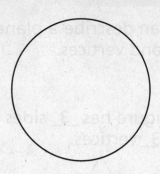

3.
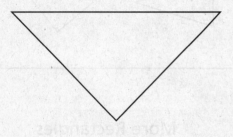

4.

5.
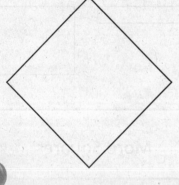

6.

Check

7. Sarah says that there are 3 squares shown below because 3 of the figures have 4 sides. Do you agree? Explain.

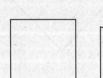 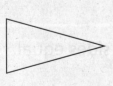 

_____

_____

Name_____

Learn the Math

You can classify a figure by the number of straight sides and the number of vertices it has.

Vocabulary

side

vertex/vertices

How many sides and vertices do these two figures have?

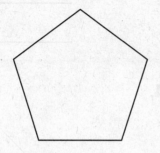
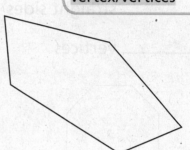

Count the straight sides.

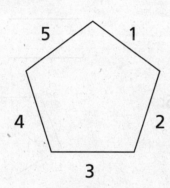
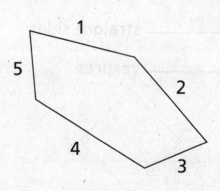

Each figure has ____ straight sides.

_____

Count the vertices.

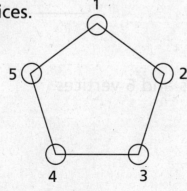
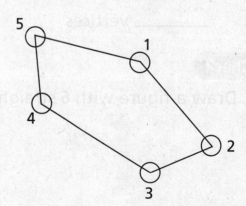

Each figure has ____ vertices.

So, the figures both have ____ straight sides and ____ vertices.

Write the number of straight sides and the number of vertices.

1.

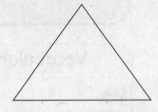

_____ straight sides

_____ vertices

2.

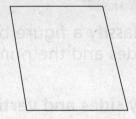

_____ straight sides

_____ vertices

3.

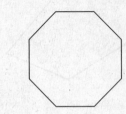

_____ straight sides

_____ vertices

4.

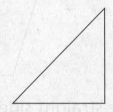

_____ straight sides

_____ vertices

5.

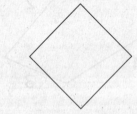

_____ straight sides

_____ vertices

6.

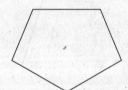

_____ straight sides

_____ vertices

7. Draw a figure with 6 straight sides and 6 vertices.

Learn the Math

You can compare figures by looking at their size and their shape.

Color the figures that are the same shape as the first figure.

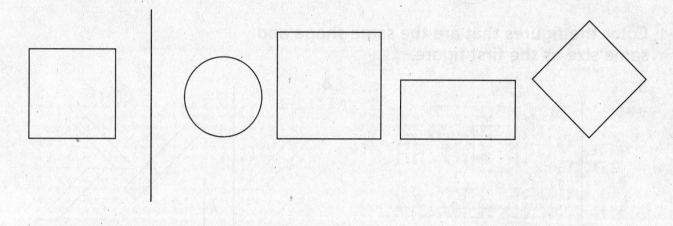

Color the figures that are the same shape and same size as the first figure.

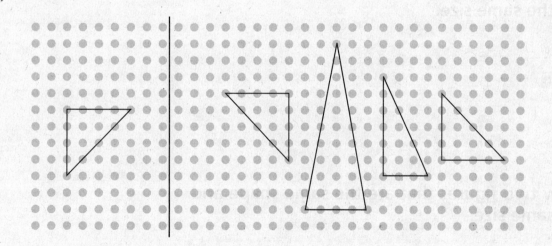

Color the figures that are the same shape as the first figure.

1.
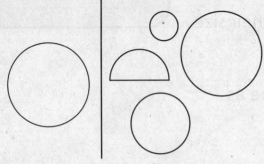

2.

Color the figures that are the same shape and same size as the first figure.

3.
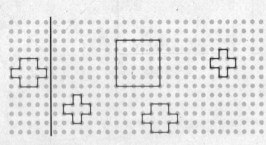

4.
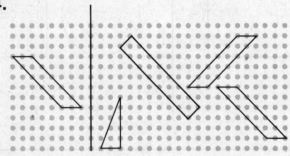

5. Draw two figures that are the same shape, but are not the same size.

6. Draw two figures that are the same shape and the same size.

Learn the Math

Congruent figures are the same size and the same shape.

These figures are congruent.

Vocabulary

congruent

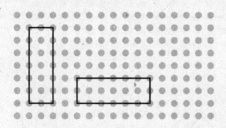

same size, same shape

These figures are not congruent.

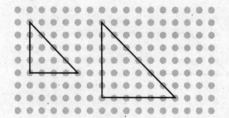

same shape, not the same size

These figures are not congruent.

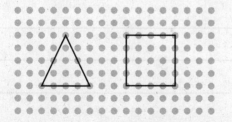

same size, not the same shape

© Houghton Mifflin Harcourt

Draw a figure that is congruent to the figure shown.

1.

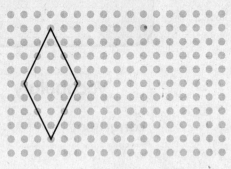

2.

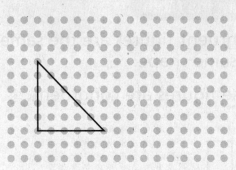

3.

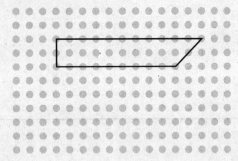

4.

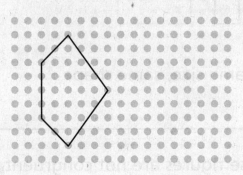

5.

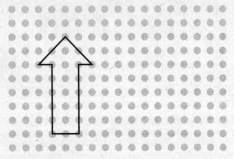

6.

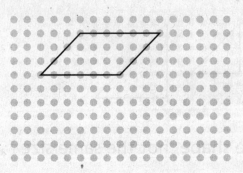

Check

7. Sherry says these figures are congruent. Do you agree? Explain.

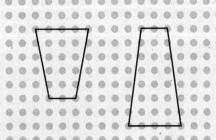

_____

_____

_____

_____

_____

NaNName_____

Learn the Math

Solid figures have length, width, and height.
They can also be called three-dimensional figures.

These are some solid figures.

This is a rectangular prism.
It is shaped like a box, with
6 flat faces.

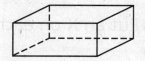

_____

This is a cube.
It is shaped like a box with
all faces equal size.

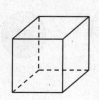

Remember: A cube is a rectangular
prism where each side is a square.

_____

This is a pyramid.
It has triangle shaped faces
that meet at a point.

_____

This is a sphere.
A sphere is shaped like a ball.

_____

This is a cone.
It has a pointed top and
sits on a flat, round face.

_____

This is a cylinder.
It is shaped like a can.

1. Circle each rectangular prism. Put an X on each pyramid.

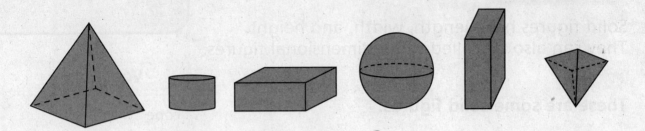

2. Circle each cylinder. Put an X on each sphere.

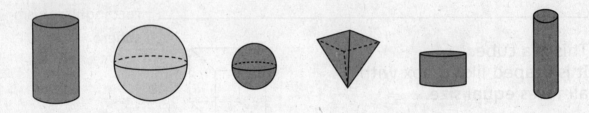

3. Circle each cone. Put an X on each cube.

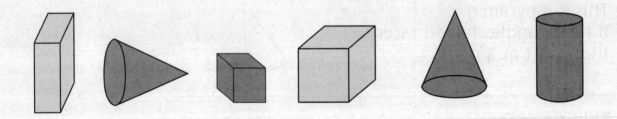

Check

7. Tino buys this box of marbles.
   He takes 4 marbles out of the box.

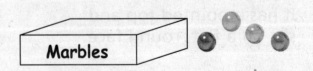

What is the shape of the box? _____

What is the shape of each marble? _____

Name_____

Learn the Math

You can describe a solid figure by the number of faces, edges, and vertices it has.

Count the number of faces.

A face is a flat surface of a solid figure.

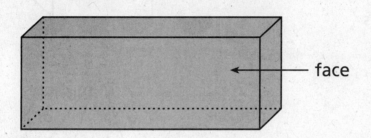

face

The figure has **6** faces.

Count the number of edges.
An edge is formed where two faces meet.

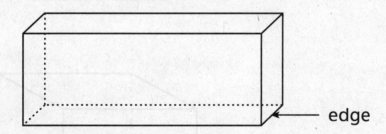

edge

The figure has **12** edges.

Count the number of vertices.
A vertex is a point where 3 or more edges meet.

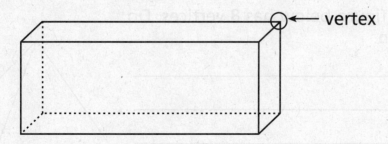

vertex

The figure has ____ vertices.

© Houghton Mifflin Harcourt

Count the faces, edges, and vertices of each
solid figure. Write how many of each.

1.

____ faces

____ edges

____ vertices

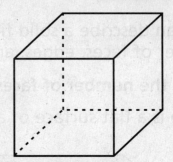

2.

____ faces

____ edges

____ vertices

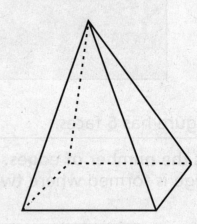

3.

____ faces

____ edges

____ vertices

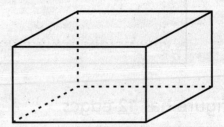

Check

4. Shannon says the figure below has 8 vertices. Do
you agree? Explain.

_____

_____

_____

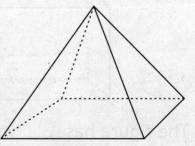

Name_____

Learn the Math

You can use nonstandard units to estimate and measure the length of an object.

Estimate the length of the crayon.

      about _____ tiles

Then measure.

Step 1 Place one tile below the left side of the crayon.

Vocabulary

length

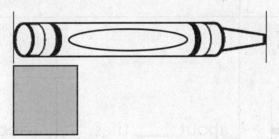

Step 2 Place more tiles to the right of the first one until you reach the right end of the crayon.

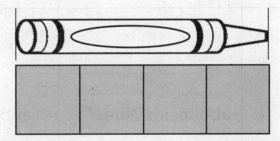

Step 3 Count the number of tiles.

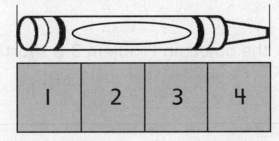

The crayon is about _____ tiles long.

Use tiles. Estimate.
Then measure.

Object	Estimate	Measure
1. 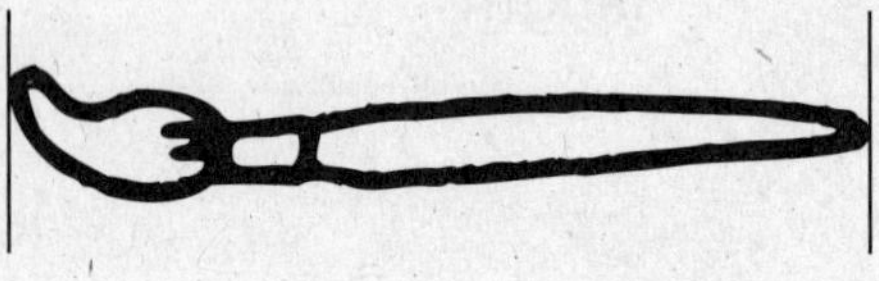	about _____ tiles	about _____ tiles
2. 	about _____ tiles	about _____ tiles
3.  Eraser	about _____ tiles	about _____ tiles

Check

4. Explain how you know the object in Problem 3 is shorter than the other two objects.

_____

_____

_____

Learn the Math

You can use inches to measure length.

The paper clip is about 1 inch long.

Vocabulary

inch

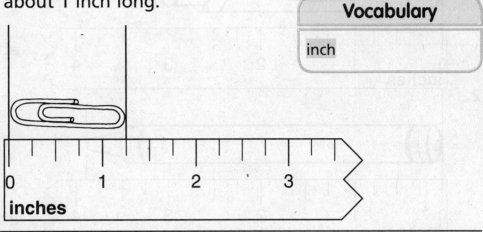

Use an inch ruler to measure the length of the eraser.

Step 1 Put the 0 mark of the ruler at the left end of the eraser.

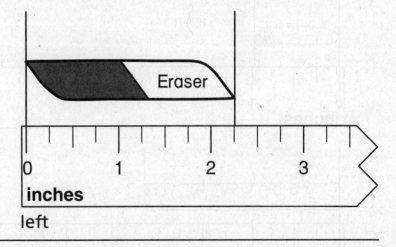

left

Step 2 Look at the right end of the eraser.

The number on the ruler closest to the right end of the eraser tells how long it is.

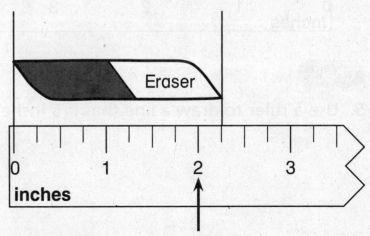

So, the eraser is about _____ inches long.

Use an inch ruler to measure the length of each object.

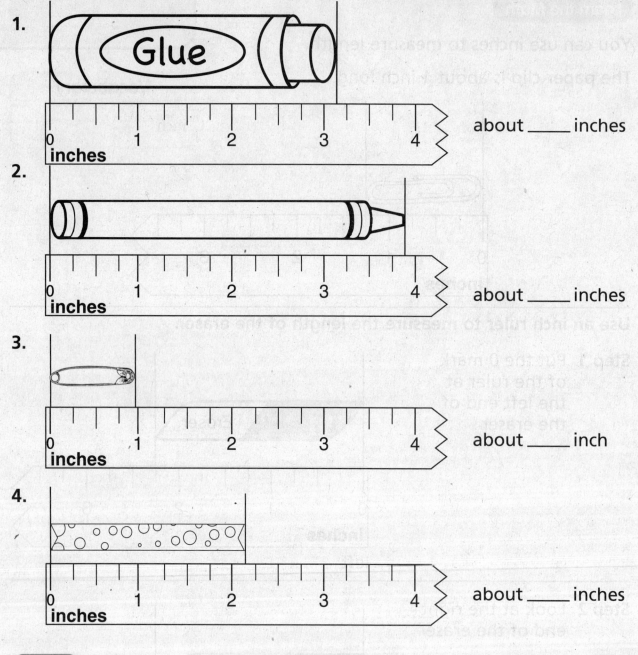

1.

about _____ inches

2.

about _____ inches

3.

about _____ inch

4.

about _____ inches

Check

5. Use a ruler to draw a line that is 5 inches long.

Name_____

Learn the Math

Capacity is the amount a container holds. Use a small unit to measure a small container. Use a large unit to measure a large container.

Vocabulary

capacity

How much can a bowl hold? Which unit is better for measuring the capacity?

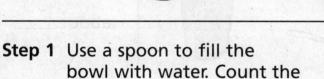

Step 1 Use a spoon to fill the bowl with water. Count the number of spoonfuls it takes.

about _____ spoonfuls

Step 2 Use a plastic cup to fill the bowl with water. Count the number of cupfuls it takes.

about _____ cupfuls

Step 3 Choose the unit that is better for measuring how much water the bowl can hold.

Which unit is better?

_____

So, the capacity of the bowl is about _____ cupfuls.

© Houghton Mifflin Harcourt

Use a spoon, cup, and real objects.
Circle the unit you would use to measure. Then measure.

Container	Unit	Measurement
1.		about _____
2.		about _____
3.		about _____

Check

4. Jennifer and Casey used a plastic cup to measure the capacity of a pot. It took Jennifer 10 cups to fill it. It took Casey only 7 cups to fill the pot. Why are the measurements different?

_____

_____

_____

Name_____

Learn the Math

Weight is the measure of how heavy an object is.

You can hold an object in your hand to help you choose a unit to measure weight. Use a light unit for a light object. Use a heavier unit for a heavier object.

Which unit is better for measuring the weight of an apple?

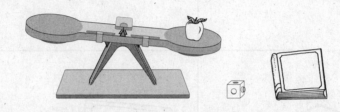

_____

Step 1 Hold the apple in your open hand.

      Think: How heavy does it feel?

_____

Step 2 Hold a book in your hand.

      Think: Is it heavier or lighter than the apple?

         _____

_____

Step 3 Hold a cube in your hand.

      Think: Is it heavier or lighter than the apple?

         _____

_____

Step 4 Choose the object that is a better unit to measure the weight of an apple.

So, the _____ is a better unit for measuring the weight of an apple.

Use a balance and real objects.
Circle the unit you would use to measure the weight.
Then measure.

Object	Unit	Measure
1.		about _____
2.		about _____
3.		about _____

Check

4. How can you use a balance to find the weight of an object?

_____

_____

_____

_____

© Houghton Mifflin Harcourt

Learn the Math

This eraser is about 1 centimeter long.

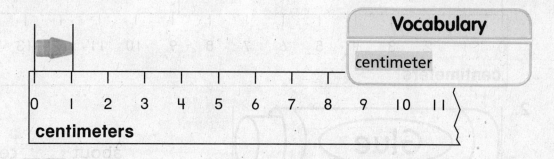

© Houghton Mifflin Harcourt

Vocabulary

centimeter

Use a centimeter ruler to measure a pencil.

Step 1 Put the 0 mark on the ruler at the left end of the pencil.

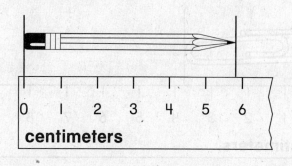

left

Step 2 Look at the right end of the pencil.

The number on the ruler closest to the right end of the pencil tells how long it is.

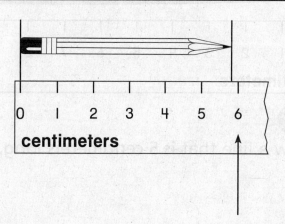

The pencil is about ____ centimeters long.

Use a centimeter ruler to measure the length of each object.

1.
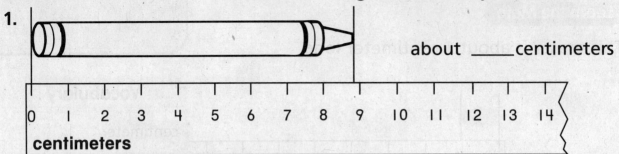

about _____ centimeters

2.
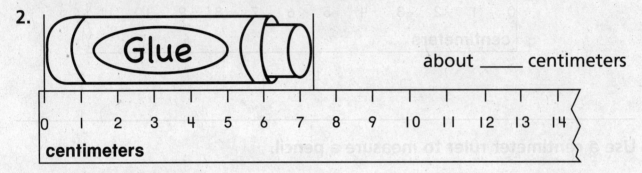

about _____ centimeters

3.
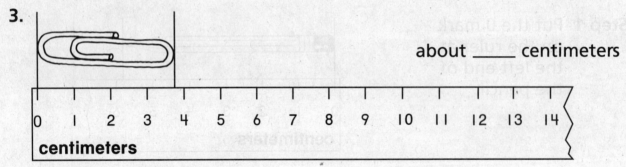

about _____ centimeters

4.
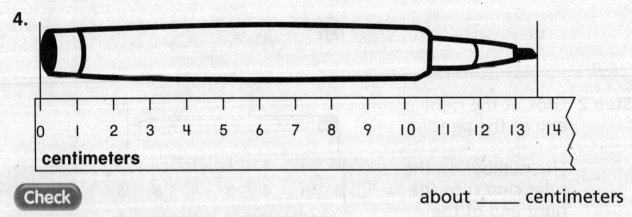

about _____ centimeters

Check

5. Draw a line that is 5 centimeters long.

Learn the Math

Perimeter is the distance around a figure.

Find the perimeter of the figure below.

Vocabulary

perimeter

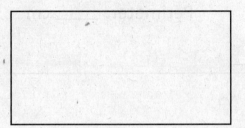

> **Remember**
>
> Another way to write centimeter is cm.

Step 1 Use a centimeter ruler to measure each side.

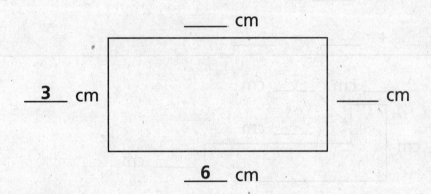

_____ cm

__3__ cm   _____ cm

__6__ cm

Step 2 Add the lengths of the sides.

__6__ + __3__ + _____ + _____ = _____

The perimeter of the figure is _____ centimeters.

Measure each side. Add to find the perimeter.

1.

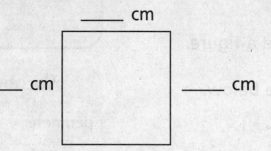

____ cm

____ cm   ____ cm

____ cm

Perimeter: ____ cm

____ + ____ + ____ + ____ = ____

2.

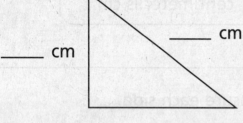

____ cm

____ cm

____ cm   Perimeter: ____ cm

____ + ____ + ____ = ____

3.

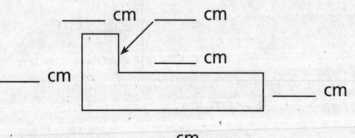

____ cm   ____ cm

____ cm

____ cm

____ cm

____ cm   Perimeter: ____ cm

____ + ____ + ____ + ____ + ____ + ____ = ____

Check

4. Draw a figure with a perimeter of 10 units. What
are the lengths of the sides?

Name____

Learn the Math

Vocabulary

area

square unit

Area is the number of square units that cover a flat surface.

You can count the square units to find the area.

1 square unit

The area of this figure is 8 square units.

Count the square units to find the area of the figure.

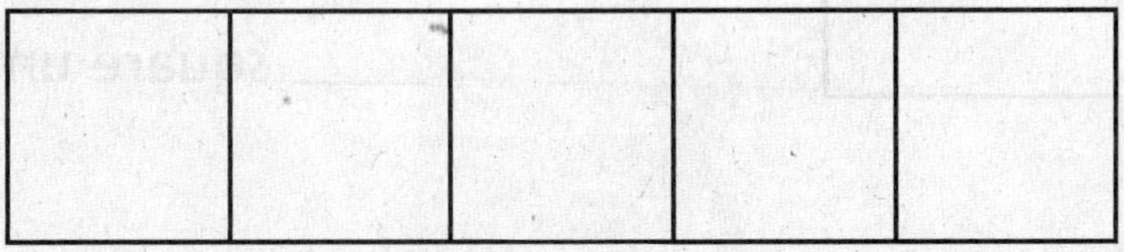

The area of the figure is ____ square units.

The area of the figure is ____ square units.

Count the square units to find the area
of the figure.

1.

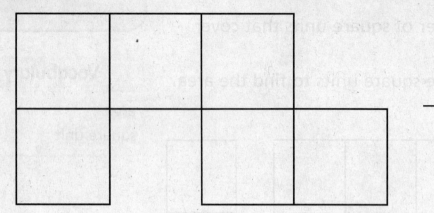

_____ square units

2.

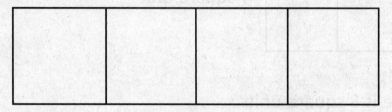

_____ square units

3.

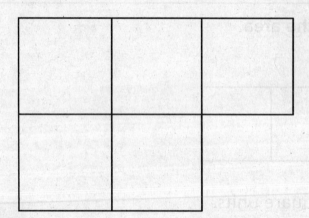

_____ square units

Check

4. Draw a figure with an area of 7 square units.

Learn the Math

You can tell if figures have equal parts.

The equal parts are the same size. The unequal parts are not the same size.

Vocabulary

equal parts
unequal parts
whole

whole

1 circle

2 equal parts

The parts are the same size.

2 unequal parts

The parts are **not** the same size.

whole

1 rectangle

___equal parts

The parts are the same size.

___ unequal parts

The parts are **not** the same size.

whole

1 square

___equal parts

The parts are the same size.

___ unequal parts

The parts are **not** the same size.

Circle the figures with equal parts.
Draw an X over the figures with unequal parts.

1.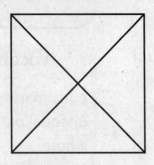

2.

3.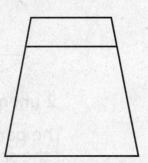

4.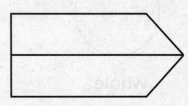

5.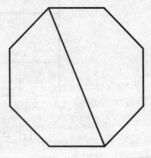

6.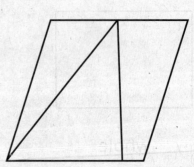

Check

7. Draw a circle with 4 equal parts. Then draw one with 4 unequal parts.

Name_____

Halves

Learn the Math

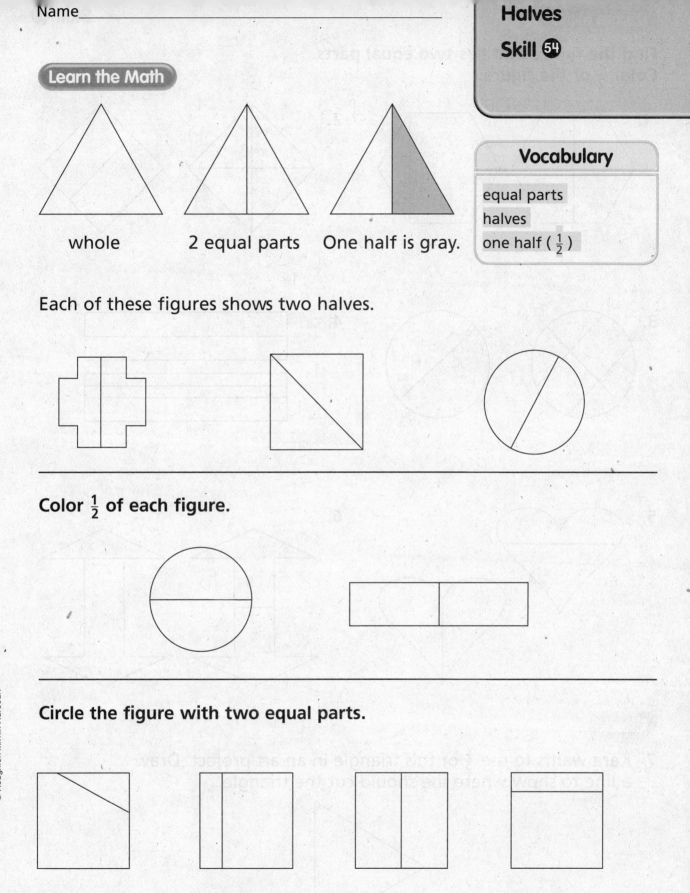

whole   2 equal parts   One half is gray.

Vocabulary

equal parts
halves
one half ($\frac{1}{2}$)

Each of these figures shows two halves.

Color $\frac{1}{2}$ of each figure.

Circle the figure with two equal parts.

© Houghton Mifflin Harcourt

Find the figure that has two equal parts.
Color $\frac{1}{2}$ of the figure.

1.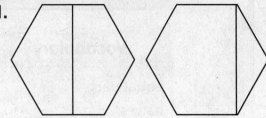

2.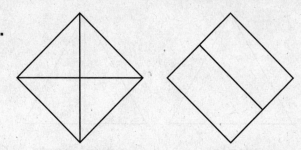

3.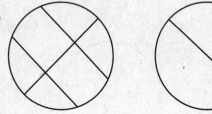

4.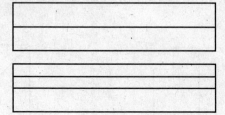

5.

6.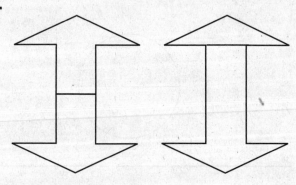

Check

7. Kara wants to use $\frac{1}{2}$ of this triangle in an art project. Draw a line to show where she should cut the triangle.

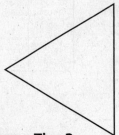

Learn the Math

1 whole	4 equal parts	4 fourths

Vocabulary

equal parts
fourths
one fourth ($\frac{1}{4}$)

1 of 4 equal
parts is gray.

One fourth
is gray.

$\frac{1}{4}$ is gray.

1 of 4 equal
parts is gray.

One fourth
is gray.

_____ is gray.

Find the figure that has four equal parts. Color $\frac{1}{4}$ of the figure.

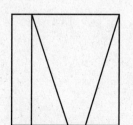 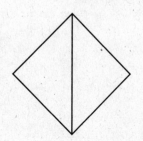 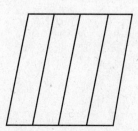 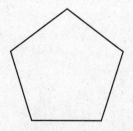

Find the figure that has four equal parts. Color $\frac{1}{4}$ of the figure.

1. 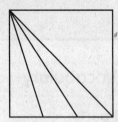   2.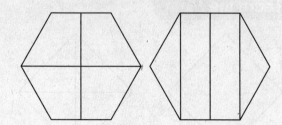

3. 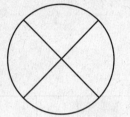   4.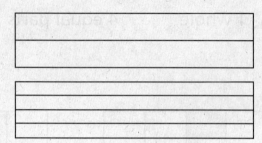

5. 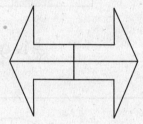   6.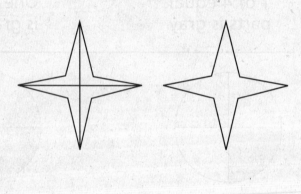

7. Martin makes a flag in the shape of a rectangle. One fourth of the flag has stars on it. Draw the flag.

Name_____

Learn the Math

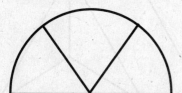 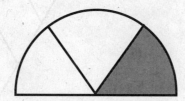

The figure has 3
equal parts.
It has 3 thirds.

One third is gray.
$\frac{1}{3}$ is gray.

Vocabulary

equal parts
one third ($\frac{1}{3}$)
thirds

Circle the figures that show thirds.

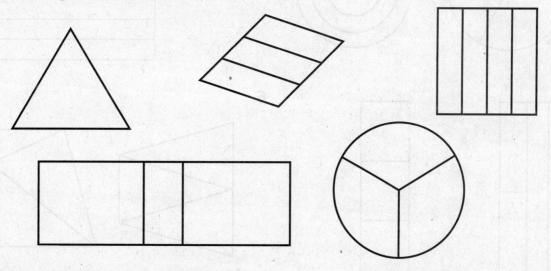

Circle the figures that are $\frac{1}{3}$ gray.

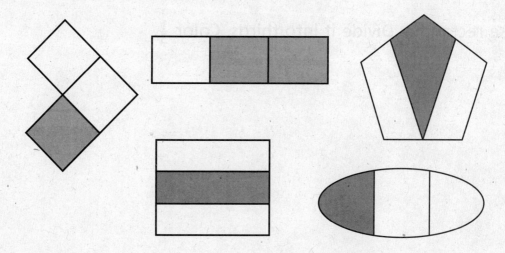

© Houghton Mifflin Harcourt

Do the Math

Find the figure that shows thirds. Color $\frac{1}{3}$ of the figure.

1.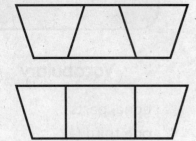

2.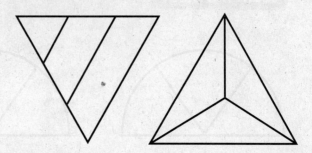

3.

4.

5.

6.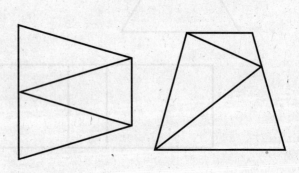

Check

7. Draw a rectangle. Divide it into thirds. Color $\frac{1}{3}$.

Name_____

Learn the Math

An event is something that happens.
You can tell whether an event is certain
or impossible.

**Is it certain or impossible that a spin will
be white?**

Vocabulary

event
certain
impossible

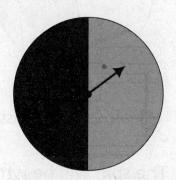

Think: There are no white sections on the spinner.

If an event will never happen, it is impossible.
You can never spin white on this spinner.

So, it is **impossible** that a spin will be white.

_____

Is it certain or impossible that a spin will be white?

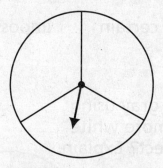

Think: There are only white sections on the spinner.

If an event will always happen, it is certain.
A spin on this spinner will always be white.

So, it is _____ that a spin will be white.

Circle *certain* or *impossible* to describe the event.

1. The spin will be gray.

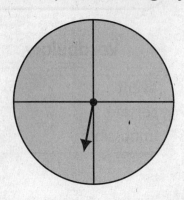

certain   impossible

2. The spin will be black.

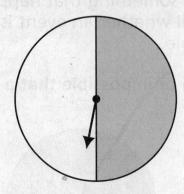

certain   impossible

3. The spin will be white.

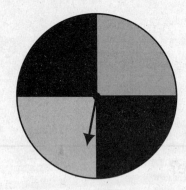

certain   impossible

4. The spin will be white.

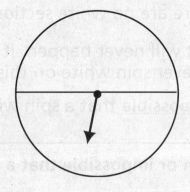

certain   impossible

Check

5. Janna says that it is impossible to spin gray using the spinner below because there are more white sections than gray sections. Is she correct? Explain.

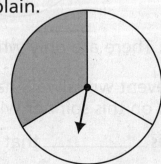

_____

_____

_____

_____

© Houghton Mifflin Harcourt